'76 0 9 5 0 0   y511

DATE DUE		
DEC 11 76		
apr 13		

Thomas Y. Crowell Company / New York

by David A. Adler

illustrated by Harvey Weiss

YOUNG MATH BOOKS

Edited by Dr. Max Beberman, Director of the Committee on
School Mathematics Projects, University of Illinois

Edited by Dorothy Bloomfield, Mathematics Specialist,
Bank Street College of Education

Library of Congress Cataloging in Publication Data Adler, David A. 3D, 2D, 1D. Written by David A. Adler and illustrated by Harvey Weiss. SUMMARY: Explains through simple experiments the principle of dimensions and how they are measured. 1. Mensuration—Juv. lit. [1. Measuring] I. Weiss, Harvey, illus. II. Title. QA465.A25 530'.8 74-5156 ISBN 0-690-00456-7
ISBN 0-690-00543-1 (lib. bdg.)

1   2   3   4   5   6   7   8   9   10

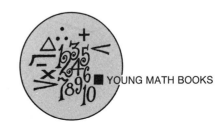

YOUNG MATH BOOKS

Perhaps you have heard people speak of
"3-D" or "three-dimensional" movies.
Three-dimensional movies *look* very real.

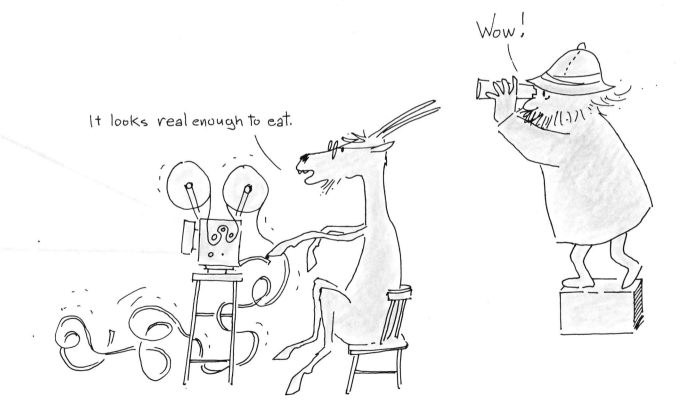

It looks real enough to eat.

Wow!

Here is an experiment that will show you what the three DIMENSIONS are.

Find five different books on a bookshelf. Some are larger than others. What makes one book large and another book small?

Stand the five books up next to each other on the bookshelf. Some books are higher than others.

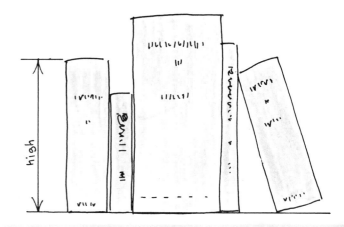

Now push the books so that each book is against the back of the bookcase. Some books are wider than others.

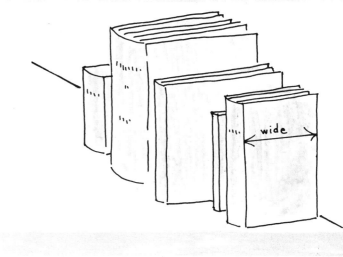

Now take the books off the bookshelf and lay each book flat on a table. Some books are thicker than others.

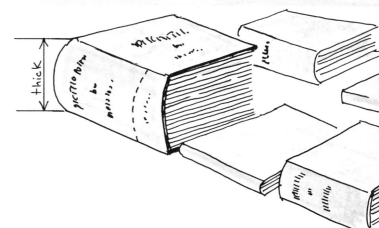

The size of a book changes when any one of its three dimensions changes: how high it is, how wide it is, and how thick it is.

A book is three dimensional.
Everything that you can pick up is three dimensional.

This page certainly has height and width. It also has thickness. Run your finger along the edge of the page. The thin edge is the thickness of the page.

This page has height, width, and thickness. It is three dimensional.

Different people give the three dimensions different names. People may say that something is tall or that it is high and mean the same thing. "Wide" and "long" also sometimes mean the same thing. So do "deep" and "thick."

No matter what words we use to talk about them, all things have three dimensions.

Find two empty boxes.
Which box is larger?

This is a big box.

Mine is bigger — I think?

OATS

This can be a hard question. What if one box
is very high and the other box is very wide?
VOLUME tells you how large a box is.

Here's something you can do to find out more about volume.

Find a lot of small toy blocks that are all the same size. If you cannot find blocks use small cans or marbles. Any things that are the same size and small will do.

How many blocks can you fit into one of the boxes you found?

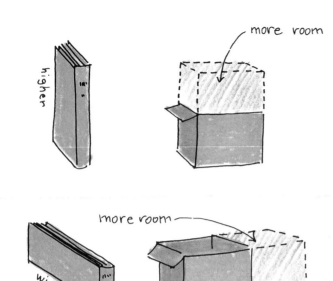

If the box were higher you could fit more inside.

If the box were wider you could fit more inside.

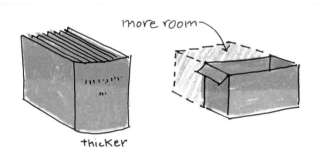

If the box were thicker you could fit more inside.

If any one of its three dimensions were larger, the box would hold more.

We call the amount a box holds, or the space that anything with three dimensions takes up, its volume.

How many blocks can you fit into the second box?

If you can fit more blocks into the second box than you can fit into the first box, the volume of the second box is larger than the volume of the first box.

36 blocks in my box.

If you can fit more blocks into the first box, then the volume of the first box is larger than the volume of the second box.

Try drawing a picture of one of the boxes.
You can show how high and how wide the
box is. It is hard to show how high, wide, and
thick the box is all in the same drawing. You are
drawing in two dimensions and you are trying to
draw something that has three dimensions. That
is one reason why it is hard to be a good artist.

If a picture is called 3-D or three dimensional it is only because the picture looks so real. The surface of the picture is still really two dimensional.

AREA is the measure of a two-dimensional surface. Here is something you can do that will show you more about area.

Have you ever tried to cover the top of a table with a tablecloth that was too small?

Try it now.

Take a small cloth or a napkin and try to cover the top of a large table.

It cannot cover the whole table. If the cloth or napkin were much longer and much wider it would cover the whole tabletop.

19

Now take some more napkins. Open them up. How many napkins must you place next to each other to cover the top of the table?

The larger the area of the tabletop the more napkins you will need.

If the table were longer you would need more napkins to cover it.

If the table were wider you would also need more napkins.

Area measures a two-dimensional surface.

Here is the floor of a room. Each tile is the same size.

How many tiles cover the floor? Count them.
How many tiles long is the floor?
How many tiles wide is the floor?

23

These are the floors of two rooms. The tiles on each floor are the same size.

Which floor has a larger area?

This shape also has area. How many squares are in this shape? The number of *whole* squares in this shape is only part of the area of the shape. Every two half squares are the same as one whole square.

What is the area of these other shapes?

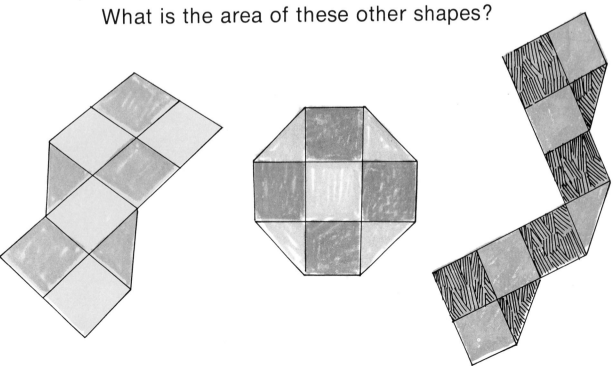

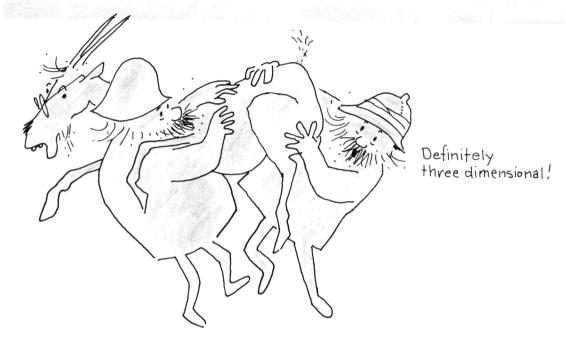

Definitely three dimensional!

Everything you can pick up is three dimensional.

Surfaces are two dimensional.
What is one dimensional?

How tall are you?

Take off your shoes. Stand with your back against the wall. Make a tiny mark with a pencil to show your height. You may need to ask someone else to make the mark. Be careful to make the mark small and to make it with a pencil so that you can erase it.

The distance from the floor to that mark is your height.

Your height is one dimensional.

Stand up straight, silly!

Your height.

How long is the longest line? To find out you can use a centimeter ruler or you can use an inch ruler.

How long is the second line?

How long is the shortest line?

The length of a line is one dimensional. It does not have width or thickness.

You can also measure this curved line.
Take a piece of string. Place one end of
the string at one end of the curved line. Now fit
the string carefully over the curved line. Cut the
string where the curved line ends.

This piece of string is the same length as the drawing of the curved line.

Straighten the string and measure it.

How far along the ruler does the string go? That is the length of the string and the length of the curved line.

The length of any line is one dimensional. It doesn't matter if it is straight or curved.

How many dimensions does a television set
have?

How many dimensions is the picture that you
see on the surface of the television screen?

How many dimensions is the height of the
television set?

Look around the room.

How many more examples of three-dimensional things, two-dimensional areas, and one-dimensional lengths can you find?

Wherever you look you can always see examples of three-dimensional things, two-dimensional areas, and one-dimensional lengths. The world is full of 3-D, 2-D, and 1-D.

David A. Adler

has been interested in mathematics since his early years in elementary school. He holds degrees from Queens College and New York University, and is presently a teacher of mathematics in the New York City school system. Mr. Adler is the author of *Base Five,* another book in the Young Math series.

Mr. Adler is married and lives in Queens, New York.

Harvey Weiss

is a sculptor whose work is in many museums and private collections. He has done a number of other things, from photography to teaching, but finds his work on children's books the most stimulating and fascinating. He is the author-illustrator of numerous books on the art of making things (from sculpture to your own books), and has also illustrated *Rubber-bands, Baseballs, and Doughnuts: A Book About Topology,* which is another title in the Young Math series.

Harvey Weiss lives and works in Greens Farms, Connecticut.